GRADE 5

M000087847

DAILY
Math Starters

180 Math Problems for Every Day of the School Year

From the Editors of DynaMath

New York • Toronto • London • Auckland • Sydney
Mexico City • New Delhi • Hong Kong • Buenos Aires

Editor: Maria L. Chang
Cover design by Michelle H. Kim
Cover photo: © Image Source/Getty Images
Interior design by Grafica, Inc.
Illustrations by thenounproject.com

ISBN: 978-1-338-15962-2

1 2 3 4 5 6 7 8 9 10   40   25 24 23 22 21 20 19 18

Table of Contents

Introduction

Welcome to *Daily Math Starters*—an easy and enjoyable way to get students revved up for learning and thinking about math every day! This collection of 180 instant skill-building math problems covers a wide range of topics, including number and operations, geometry, measurement, data analysis, and algebra. The problems in this book have been carefully selected to ensure they are developmentally appropriate yet increase in challenge as the weeks go by. Each week's set of problems features a variety of skills and problem types so students get a continuous spiral review of math concepts and standards throughout the year.

But don't think you have to present the problems in the order they appear in the book. Pick and choose a problem to suit your needs. You can use any of these math problems to preview the day's lesson or review a recent one. Here are a few other ways you can use them in your classroom:

- **Do-Now Activity**—Set students working on a problem as soon as they arrive in the morning to help them get settled and ready for learning. Choose a problem and make enough photocopies for everyone. Stack the copies on a table next to the door so students can pick one up as they enter. Alternatively, you can write or display the problem on the board and have students record their work and solution on a blank sheet of paper or in their math journal. After five minutes, invite students to share their answers and discuss the problem as a class.

- **Homework Packet**—Photocopy a week's worth of problems and send them home on Monday to be returned on Friday. This allows students to complete the problems at their own pace.

- **Seatwork for Fast Finishers**—Keep a file folder filled with copies of problems for those times when you need a quick activity to keep your fast finishers occupied. They can work on the problems individually or with a partner.

- **Exit Ticket**—Just as you can use these math problems to start the day, you can also use them to assess what students have learned at the end of the day. Select a problem related to a lesson you've just taught and have students submit their solution on their way out the door. Make sure they show how they solved the problem and not just write the answer. You can also let students know that if they're still unsure about the concept, they can write any questions they have on the "ticket."

- **Whole-Class Quiz Game**—Use the problems in this book to put together an end-of-the-month quiz game. The problems for each week are designed to go from easy to challenging and can be sorted into key math topics. Divide the class into small groups to solve the problems.

Any way you choose to present this collection of problems, your students are sure to come out winners as they get more practice and hone their math skills every day of the school year. Enjoy!

1

Name four units of length in the customary system of measurement.

_____   _____

_____   _____

Measurement

2

In the number 12,567,908, which digit is in the thousands place? What number does that digit represent?

Place Value

3

Unscramble these math terms.

oeaqntiu   _____

nddresuh   _____

acimeld   _____

Math Vocabulary

4

Put the following numbers in order from smallest to largest.

2.10   2.05   3.01   2.15

_____   _____   _____   _____

Ordering Decimals

5

Jake earned $5 a day every day in August for walking his neighbor's dog. How much money did Jake earn in total?

Money Word Problem

6

Penny is to *dime*

as *dime* is to _____ .

Logical Reasoning & Math Vocabulary

7

Which place value is 100 times as great as the hundreds place?

Place Value

8

Fill in the missing digits in this equation.

$$
\begin{array}{r}
1,\_\,9\ 4 \\
+\ 5\ \_\ \_ \\
\hline
1,\,9\ 8\ 9
\end{array}
$$

Addition: Missing Digits

9

Of these decimal numbers, which is the smallest? Explain why.

0.75   0.075   7.50

10

The digits of 565, 1,221, and 5,225 read the same left to right and right to left. Multiply each one by 2. Which product also reads the same left to right and right to left?

11

Which shape does not belong? Why?

rectangle   pentagon   triangle

cone   hexagon

Two- and Three-Dimensional Figures

12

Solve.

$$10 \times 100 \times 1,000 \times 10,000 \times 0 = \underline{\hspace{2cm}}$$

Multiplying Numbers Ending in 0s

13

Would you rather have a nickel a day for the month of February, or a dime a month for a whole year? Explain your answer.

Money

14

How many quists equal 1 quiggle?

$$1 \text{ quiggle} = 2 \text{ quirks}$$
$$1 \text{ quirk} = 3 \text{ quacks}$$
$$2 \text{ quacks} = 1 \text{ quist}$$

Algebraic Thinking

15

Say there are 175 school days left until summer vacation. How many 5-day weeks is that?

Division Word Problem

16

What is the sum of the digits on a phone keypad?

17

What is the product of the digits on a phone keypad? Estimate first! Is it greater than or less than 1,000,000?

18

What fraction of a dollar is 80 cents? Simplify your answer.

19

Evaluate the expression 437 ÷ 23. Show your work.

Dividing 3-Digit Numbers by 2-Digit Numbers

20

In a room of cats and chickens, there are exactly 18 legs. There are 6 animals in all. How many are chickens and how many are cats?

Multistep Word Problem

Name _____

21

Solve.

$$\frac{2}{5} + \; ? \; = \frac{9}{10}$$

Adding Fractions With Unlike Denominators

22

True or false? Explain your answer.

$$10 \times 10 \times 10 \times 10 \times 10 = 1{,}000 \times 10 \times 10$$

❏ True   ❏ False

Multiplying Numbers Ending in 0s

23

If $n = 5$, what is the value of $n \times n + n$?

Evaluating Variable Expressions

24

Which three numbers below share a common factor besides 1?
What is their common factor?

11   21   31   41   51   61   71   81

Factors

25

Gracie ate $\frac{1}{2}$ as many grapes as Rachael. Rachael ate $\frac{1}{4}$ as many as Priya. Priya ate 16 grapes. How many grapes did the others eat?

Multistep Word Problem

26

What number is 10,000 times as much as 45?

Multiplying Numbers Ending in 0s

27

Which pair of numbers does not fit? Why?

5	15
6	18
4	12
7	18

Algebraic Thinking

28

What are the next two numbers in this pattern?

3, 6, 12, 21, 33, _____, _____

Number Patterns

29

A speedy ice-skater can spin in a circle 4 times per second.
How many spins is that in 1 minute?

Multistep Word Problem

30

A chef made $3\frac{1}{2}$ dozen cupcakes for Elena's birthday party.
How many cupcakes did the chef make?

Fraction Word Problem

31

What whole number does each letter represent?

$$F > P$$
$$P > S$$
and
$$F + P + S = 6.$$

F = _____   P = _____   S = _____

Logical Reasoning

32

Which number is greater: 6 tenths or 55 hundredths?
Use a diagram to prove your answer.

Comparing Decimals

33

What number multiplied by itself has a product of one less than 50?

Mixed Operations

34

Use an area model with partial products to solve 127 × 28.

Multiplying 3-Digit Numbers by 2-Digit Numbers

35

It takes Hansel 4 seconds to dip an apple into caramel.
How many caramel apples can he dip in 2 minutes?

Multistep Word Problem

36

Where would you put 6 in this Venn diagram? Why?

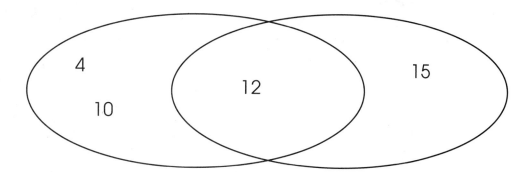

37

What is 10 more than twice the product of 6 × 5?

38

What fraction of the names of the months begins with a vowel?

39

A plane flies at a speed of 535 miles per hour for 3 hours. How far does the plane travel?

Multiplication Word Problem

40

If you save a quarter every day for one year, how much will you save? (It's not a leap year.)

Money Word Problem

41

What is $\frac{1}{5}$ of 30?

Multiplying Fractions by Whole Numbers

42

Does the following array represent a multiplication or a division problem? Explain.

Logical Reasoning

43

My tenths place is twice my hundredths. My tens is twice my ones.
My tenths and tens are the same. My ones place is 3.
What number am I?

_____ _____ . _____ _____

Logical Reasoning

44

The LCM (lowest common multiple) of two numbers is 24.
The difference between the numbers is 4.
What are the two numbers?

Logical Reasoning

45

Corey's Cookie Factory can make 200 cookies per minute.
How many cookies can the factory make in 24 hours?

Multistep Word Problem

46

What is the value of *p* in this equation?

$$21 \times p = 14 \times 9$$

Evaluating Variable Expressions

47

Express this number in standard form and as an exponent.

$$10 \times 10 \times 10 \times 10 \times 10 \times 10$$

Standard Forms & Exponents

48

What is the fewest number of small cubes you would need to construct a larger cube?

Three-Dimensional Figures

49

Write the number sentence in fraction form.

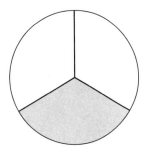 +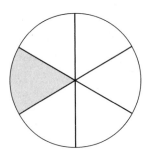

Writing Fraction Number Sentences

50

Supermarket shoppers get a free turkey once they spend $300. The Gobbler family spends $78 each week. How many weeks will it take them to earn a free turkey?

Multistep Word Problem

51

Use the numbers in the top row to find the missing digit
in the bottom row.

2	5	3	6	5	5
1	0	1	8	___	5

Logical Reasoning

52

Which two of the fractions below have the sum of $\frac{7}{12}$?

$$\frac{1}{6} \qquad \frac{1}{5} \qquad \frac{1}{4} \qquad \frac{1}{3} \qquad \frac{1}{2}$$

Adding Fractions With Unlike Denominators

53

If 1 gallon of paint covers 200 square feet,
how much paint is needed to cover 500 square feet?

Measurement Word Problem

54

Place parentheses in the following expression to make it equal to 84.

$$4 \times 12 + 36 \div 9 - 8 = 84$$

Mixed Operations

55

The *Mayflower*, the ship the Pilgrims sailed on, was 110 feet long. How long was the ship in inches?

Measurement Word Problem

56

Use the following digits and symbols to create an expression that is equal to 4.3. (Hint: There are two possible answers.)

1   2   5   8   .   .   +

57

Write an equation that represents the subtraction problem below.

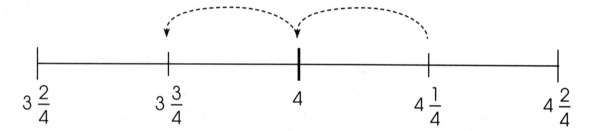

$3\frac{2}{4}$ $3\frac{3}{4}$ 4 $4\frac{1}{4}$ $4\frac{2}{4}$

58

Which of the following fractions are equivalent?

$\frac{2}{8}$ $\frac{3}{4}$ $\frac{9}{15}$ $\frac{4}{16}$ $\frac{12}{32}$

59

Corn muffins are sold 12 to a box. Mike needs 75 corn muffins.
How many boxes does he need to buy?

Multistep Word Problem

60

Which is longer: 3 yards and 15 inches or 150 inches?
How much longer?

Measurement

61

How many more hours are there in December than in November?

Time Word Problem

62

What is the missing denominator?

$$\frac{2}{5} + \frac{1}{x} = \frac{11}{15}$$

Adding Fractions With Unlike Denominators

63

Solve.

$$\frac{3}{4} \times 20 =$$

Multiplying Fractions by Whole Numbers

64

On Monday, Jayden built a 22.4-inch-tall snowman.
By Tuesday, it had melted down to 18.6 inches tall.
By how many inches did the snowman melt?

Decimal Word Problem

65

Out of 16 aliens, some had 2 arms. The rest had 4 arms.
In all, there were 44 arms. How many aliens had 2 arms
and how many had 4?

Multistep Word Problem

66

Solve. Simplify your answer.

$$5 \times \frac{2}{15} =$$

Multiplying Fractions by Whole Numbers

67

Solve for *y*.

$$185 - y = 67 \times 2$$

Evaluating Variable Expressions

68

It takes 8 tomatoes to make a jar of sauce. How many tomatoes would it take to make $3\frac{1}{2}$ jars of sauce?

Fraction Word Problem

69

Lindy has 3 times as many grapes as carrots. She has 19 more carrots than crackers. She has 25 crackers. How many grapes does Lindy have?

Multistep Word Problem

70

John receives $14.20 per hour. How much would John earn for working 40 hours?

Money Word Problem

71

Which two of the fractions below have a difference of $\frac{1}{4}$?

$$\frac{6}{8} \qquad \frac{1}{2} \qquad \frac{1}{3} \qquad \frac{4}{6} \qquad \frac{1}{7}$$

Subtracting Fractions With Unlike Denominators

72

Solve.

$$\frac{1}{4} \div \frac{1}{8} =$$

Dividing Two Fractions

73

What are the next three numbers in this pattern?

72, 36, 40, 20, 24, _____, _____, _____

Number Patterns

74

Suppose 12 polar bears weigh a total of 6 tons, and each bear is the same weight. How many pounds does each bear weigh?

(Hint: 1 ton = 2,000 pounds)

Multistep Word Problem

75

Berta the cat meows every 9 minutes. Percy the cat meows every 12 minutes. How many minutes will it take for them to meow at the same time?

Multistep Word Problem

76

Which place value is 1,000 times greater than the hundredths place?

Place Value

77

Find the number being described.

- It's between 1 and 20.
- It has exactly 3 factors.
- The sum of its factors is 13.

Logical Reasoning

78

Car 1 goes 25 mph for 7 hours. Car 2 goes 60 mph for 3 hours. Which car goes farther? By how many miles?

Multistep Word Problem

79

Four out of every 5 monkeys at the zoo like to eat grapes. There are 40 monkeys at the zoo. How many monkeys like grapes?

Multistep Word Problem

80

The sum of the ages of JJ, AJ, and their son TJ is 79. AJ is 6 years older than JJ. All of their ages are prime numbers. What are their ages?

Logical Reasoning

81

In the number 128.98, which digit is in the thousandths place?

Place Value

82

A student wrote the following equation for the product 27 × 6.
Is he correct? Explain.

$$27 \times 6 = (20 \times 6) + (70 \times 6)$$

Properties of Multiplication

83

A light bulb can burn for 6,000 hours. How many days is that?

Time Word Problem

84

Nine shoes that are each 8 inches long are lined up heel to toe.
How many 6-inch shoes would it take to equal the same length?

Multistep Word Problem

85

Bob the Baker says that $\frac{6}{10}$ of each of his muffins is flour.
One muffin weighs 0.8 ounces. How much of that is flour?

Multistep Word Problem

86

Which number has more factors: 9 or 19?

Factors

87

True or false? Explain your answer.

$$\frac{6}{8} = \frac{2}{3}$$

❏ True   ❏ False

Equivalent Fractions

88

Solve for *x*.

$$5 \times 5 > 20 + x$$
$$20 + x > 92 \div 4$$

Evaluating Variable Expressions

89

Heather can bake 12 cupcakes at a time. She needs 65 cupcakes.
How many batches must she make?

Division Word Problem

90

Jaden's car can travel 30 miles on one gallon of gas.
A gallon of gas costs $2.49. How much will it cost for him
to travel 1,800 miles?

Multistep Word Problem

91

My tens digit is twice my hundreds. My ones digit is 5 more than my tenths. My tenths and tens are both 2. What number am I?

_____ _____ _____ . _____

Logical Reasoning

92

What is $\frac{2}{3}$ of 60? Write an equation and solve it.

Multiplying Fractions by Whole Numbers

93

Find the missing digit to make this equation true.
Use the same digit for each blank.

_____ × 3 − 12 = _____

Mixed Operations

94

Kim has twice as many frogs as Ed has.

Ed has 4 more frogs than Tina has. Tina has 14 frogs.

How many frogs in total do the three friends have?

Multistep Word Problem

95

Paul's favorite movie is 1 hour, 22 minutes long.

How many times can he watch it in an 8-hour period?

How much time will he have left over?

Multistep Word Problem

96

Find the sum. Express your answer as an improper fraction and a mixed fraction.

$$\frac{6}{5} + \frac{8}{5} =$$

Improper Fractions & Mixed Numbers

97

Write your own equation with a variable based on the following tape diagram.

123.6

| 76.9 | z |

Writing Variable Expressions

98

What is the sum of all the factors of the following numbers?

1   6   11   15

Factors

99

Use partial products or draw an area model to solve
the following problem.

$$3,604 \div 53$$

Dividing Multi-Digit Numbers by 2-Digit Numbers

100

A Valentine's chocolate box had 30 pieces. On Monday, Sara ate $\frac{1}{3}$.
On Tuesday, Kayla ate $\frac{1}{5}$ of the remaining pieces. How many are left?

Multistep Word Problem

101

Order these decimals, from greatest to least.

3.997   3.99   0.9   3.999

_____   _____   _____   _____

Ordering Decimals

102

Remove two lines so that only two squares remain with no extra lines.

Logical Reasoning

103

True or false?

$3,000 \times 50 = 50,000 \times 3 = 300 \times 500$

❏ True   ❏ False

Multiplying Numbers Ending in 0s

104

A snail can travel about 0.03 miles per hour (mph).
A giant tortoise can travel about 0.17 mph.
How many mph faster is the tortoise?

Decimals Word Problem

105

A recipe calls for 6 cups of barbecue sauce. Derek has only
a $\frac{1}{3}$-cup measuring cup. How many times must he fill it
to measure out 6 cups?

Measurement Word Problem

106

If 5 snickels = $\frac{1}{4}$ of a snerd, how many snickels = $\frac{1}{2}$ of a snerd?

Logical Reasoning

107

What is the greatest common factor of these three numbers?

27   54   63

Greatest Common Factor

108

Which one is true?

$(6 \times 3) + 7 = 5 \times 5$

$(2 + 6) \times 2 = (6 \times 2) + 2$

$(5 - 5) \times 5 = (5 \times 5) - 5$

Mixed Operations

109

In a frog-jumping contest, Croaky jumps $3\frac{1}{16}$ feet.
Fluffy jumps $4\frac{1}{8}$ feet. How much farther did Fluffy jump?

Fraction Word Problem

110

Roy is twice as old as his sister. The sum of their ages is $\frac{1}{2}$ their mom's age. Their mom is 36 years old. How old are Roy and his sister?

Logical Reasoning

111

Continue the pattern.

$$\frac{1}{2}, \ \frac{2}{4}, \ \frac{4}{8}, \ \frac{8}{16}, \ \underline{\hspace{1.5cm}}, \ \underline{\hspace{1.5cm}}, \ \underline{\hspace{1.5cm}}$$

Number Patterns

112

Name the two numbers described below.

- Their greatest common factor is 7.
- Neither is a multiple of the other.
- Both are less than 25.

Logical Reasoning

113

Find the value of *q*.

$$8 \times q = 144$$

Evaluating Variable Expressions

114

A skydiver falls 120 miles per hour until her chute opens.
At that speed, how long will it take to fall 1 mile?

Multistep Word Problem

115

A cat eats $\frac{1}{4}$ ounce of treats each day. How many days will it take
the cat to eat 1 pound of treats? (Hint: 1 pound = 16 ounces)

Multistep Word Problem

116

Use the following digits to create an improper fraction and
its mixed-number equivalent.

$$2 \qquad 2 \qquad 3 \qquad 3 \qquad 8$$

Improper Fractions & Mixed Numbers

117

Put decimal points in the addends so that the equation is correct.

$$403 + 44 + 35 = 8.78$$

Adding Decimals

118

What is the difference between the formula for the perimeter
of a rectangle and the formula for the area of a rectangle?

Area & Perimeter

119

In a box of 120 crayons, 8 are a shade of yellow.
Write three different fractions that express the portion
of the crayons that are yellow.

Equivalent Fractions

120

The per-gallon price of gas dropped from $2.16 one week to
$1.97 the next. How much more did 15 gallons cost the first week
than the second?

Multistep Word Problems

121

Complete this statement.

2 is to 4

as 5 is to 10

as 12 is to 24

as _____ is to $\frac{1}{2}$.

122

What is the decimal equivalent of $\frac{3}{5}$? Represent your answer in decimal form along with a base-10 grid.

123

How many tiddlywinks are in 7 whispers?

1 whisper = 4.3 tiddlywinks

124

Mike's Munchy Mix serves 6 people and has $\frac{3}{4}$ of a cup of walnuts. How many cups of walnuts would he need to serve 24 people?

Multistep Word Problem

125

If you are given a 24-inch piece of string, what size quadrilateral could you make with it that will give you the greatest area?

Area Word Problem

126

What is 0.875 written as a fraction?

Converting Decimals to Fractions

127

Solve.

$$\frac{1}{4} \div \frac{1}{2} =$$

Dividing Two Fractions

128

If you wrote the number 100 1,000 times, how many zeroes would you have written?

Number Sense

129

Wizard Will's wand is 9.24 inches long. His big sister, Wilma, has a wand that's 12.37 inches long. How much longer is Wilma's wand than Will's?

Decimal Word Problem

130

What is the area of the wall below, not including the window?

2 ft

2 ft

10 ft

15 ft

Area

131

Fill in the missing number to make this equation true.

$$2.7 = 0.27 \times \underline{\hspace{2cm}}$$

Multiplying Decimals

132

Fill in the missing numerator and denominator.

$$\frac{8}{100} = \frac{}{50} = \frac{16}{}$$

Equivalent Fractions

133

What are the dimensions of a square where the perimeter equals the square units in the area?

Area & Perimeter

134

Alexis had a 7-inch pencil. She sharpened off $\frac{1}{4}$ inch each day. How many days did the pencil last?

Fraction Word Problem

135

A single honeybee has 4 wings and 6 legs. In a hive of 25,000 honeybees, how many more legs than wings are there?

Multistep Word Problem

136

Write the answer to this expression in standard form.

1 ten thousand 6 tens − 6 thousands 9 hundreds + 3 tens 5 ones

Place Value

137

Fill in the blanks.

40 is _____ times _____ than 0.004.

Place Value

138

Which two fractions below have a sum of $\frac{11}{12}$?

$\frac{5}{12}$ $\frac{3}{4}$ $\frac{1}{2}$ $\frac{5}{6}$

Adding Fractions With Unlike Denominators

139

What is this rectangle's length?

l = _____ cm

Area = 80.6 cm²

w = 6.5 cm

Area

140

Melody worked as a babysitter 4 hours a day, 2 days a week for 9 weeks. She made a total of $460.80. How much was she paid per hour?

Multistep Word Problem

Name _____

141

On the number line below, label the point that equals $\frac{2}{3}$.

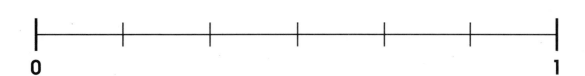

0   1

Identifying Fractions

142

Find the product. Simplify your answer.

$$\frac{2}{9} \times \frac{3}{6} =$$

Multiplying Two Fractions

143

Find the product of 8,469 and 1.7 without using a calculator or performing a computation.

$$8,469 \times 1,700 = 14,397,300$$
$$8,469 \times 170 = 1,439,730$$
$$8,469 \times 1.7 =$$

Number Patterns

144

Draw a line segment from point (3,1) to point (2,0).

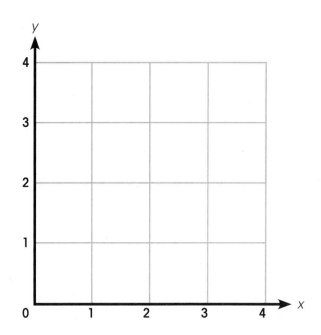

Coordinate Plane

145

The perimeter of a rectangle is 30 cm. The length is 4 times as great as its width. What are the measurements of the sides?

Perimeter Word Problem

146

Fill in the next two numbers in the pattern. What is the pattern?

3, 6, 12, 15, 30, 33, 66, 69, _____, _____

Number Patterns

147

What is the mystery fraction?

- It is equivalent to $\frac{1}{2}$.

- Its denominator is 8 greater than its numerator.

Logical Reasoning

148

Which number isn't a perfect square? Explain your answer.

4   16   26   36   49

Squares

149

January 3rd is the third day of the year. What day of the year is Earth Day, April 22nd? (Note: Assume it is not a leap year.)

Time Word Problem

150

Amy and Timmy are sharing a cookie. Amy ate $\frac{2}{3}$ of the cookie. Timmy ate $\frac{1}{8}$ of the cookie. What fraction of the cookie did they eat altogether?

Fraction Word Problem

Name _____

151

What is $\frac{19}{20}$ written as a decimal?

Converting Fractions to Decimals

152

What factor, other than 1, is shared by these numbers?

34   51   102

Factors

153

What is the formula for the volume of a cube?

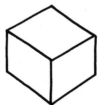

Volume

154

Plot the following points on a coordinate grid: (3,2), (3,5), (8,2), and (8,5). If you connect the points, what type of polygon does it form?

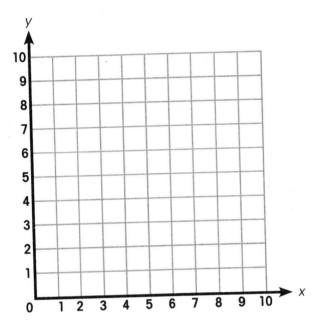

Coordinate Plane

155

A combination of 9 squares and triangles have a total of 29 corners. How many are squares and how many are triangles?

Multistep Word Problem

Name _____

156

Solve.

$$\text{January } 6 = \frac{1}{6} \qquad \text{February } 3 = \frac{2}{3}$$

$$\text{January } 6 + \text{February } 3 = \text{what date?}$$

Logical Reasoning

157

Find the product.

$$\frac{3}{5} \times 15 =$$

Multiplying Fractions by Whole Numbers

158

On a deep-sea mission, Trey starts at the coordinates (1,2) and ends at (6,2). How many units along the x-axis did Trey travel?

Coordinate Plane

159

Use partial quotients to solve 7,488 ÷ 117.

Dividing Multi-Digit Numbers by 3-Digit Numbers

160

One quart of lemonade calls for $\frac{1}{4}$ cup of sugar.
Kiara wants to make 2 gallons of lemonade for her class.
How much sugar does she need?

Multistep Word Problem

161

Solve.

$$3\frac{1}{4} - 2\frac{3}{8} =$$

Subtracting Mixed Numbers With Unlike Denominators

162

Did the student subtract correctly? Explain why or why not.

$$\begin{array}{r} 36.43 \\ -\ 34.6 \\ \hline 32.97 \end{array}$$

Subtracting Decimals

163

Fill in the area model with partial products and solve.

	40	5
30		
7		

Multiplying 2-Digit Numbers by 2-Digit Numbers

164

Plot the following points on a coordinate grid: (2,3), (4,6), (7,3), and (9,6). If you connect the points in a shape, what type of polygon do you get?

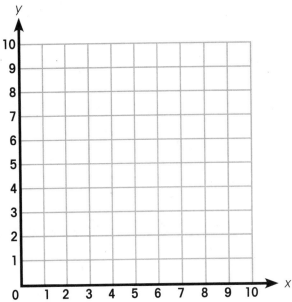

Coordinate Plane

165

What is the volume of a sand castle with a width of 8 feet, a length of 11 feet, and a height of 4 feet?

Volume

Name _____

166

Fill in the missing numerator.

$$\frac{7}{25} = \frac{}{125}$$

167

Use this set of numbers and a decimal point to create the smallest number possible.

0   9   0   7   2

168

Continue the pattern.

1, 9, 25, 49, 81, _____, _____

169

Kelly rode her bike for 5 miles at a speed of 15 mph.
How long did she spend riding her bike?

Multistep Word Problem

170

Area of large square = 16 sq. cm.
What's the area of one of the small triangles?

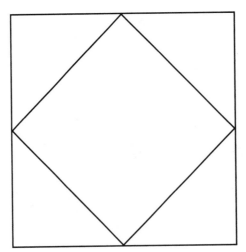

Area

171

Remove two numbers so that the sum of the remaining numbers is 55.

3   15   17   37   42

Adding Three Numbers

172

Which prime number(s) is (or are) evenly divisible by 7?

Prime Numbers

173

What is wrong with this sentence?

A rectangular prism has a height of 4 cm, a length of 5 cm, and a volume of 60 square cm.

Measurement

174

Take a square with 5-inch sides. Double the length of each side.
How many times as great is the second square's area as the first's?

Area

175

There are 24 frames of film shown per second of a movie.
How many hours and minutes will it take to show 90,720 frames?

Multistep Word Problem

176

Which two fractions below have a sum of exactly $\frac{3}{8}$?

$$\frac{1}{4} \qquad \frac{2}{5} \qquad \frac{3}{17} \qquad \frac{3}{24}$$

Adding Fractions With Unlike Denominators

177

Radius is to *diameter* as

_____ is to *whole*.

Logical Reasoning & Math Vocabulary

178

If you cut straight through this cube along the lines, how many smaller cubes will you have?

Volume

179

A student attends 180 days of school each year beginning in kindergarten. If he has perfect attendance, in what grade will he attend her 1,000th day?

Multistep Word Problem

180

In a go-cart race, you are in the lead. Ted is behind Mel and Bill. Mel is in front of Bill. Who is in first place?

Logical Reasoning

Answer Key

WEEK 1 (pages 5–6)
1. Possible answers: inch, foot, yard, mile
2. 7; 7,000
3. equation; hundreds; decimal
4. 2.05, 2.10, 2.15, 3.01
5. $155

WEEK 2 (pages 7–8)
6. dollar
7. ten thousands place
8. 1,**394** + 5**95** = 1,989
9. 0.075; when comparing place value, 0.075 is smaller than the others in the ones and tenths place.
10. 2,442 (1,221 × 2)

WEEK 3 (pages 9–10)
11. cone; it's a 3-dimensional shape, while the others are 2-dimensional
12. 0
13. A nickel a day for the month of February yields $1.40, as opposed to $1.20 for a dime a month for a whole year.
14. 3 quists
15. 35 weeks

WEEK 4 (pages 11–12)
16. 45
17. 0; less than 1,000,000
18. 4/5
19. 19
20. 3 chickens and 3 cats

WEEK 5 (pages 13–14)
21. 5/10 or 1/2
22. True; both sides of the equation equal 100,000.
23. 30
24. 21, 51, and 81; 3
25. Gracie ate 2 grapes, and Rachael ate 4.

WEEK 6 (pages 15–16)
26. 450,000
27. 7 and 18; the other pairs follow an $a \times 3 = b$ pattern.
28. 48, 66
29. 240 spins
30. 42 cupcakes

WEEK 7 (pages 17–18)
31. F = 3, P = 2, S = 1
32. 6 tenths; diagrams will vary.
33. 7
34.

	100	20	7
20	2,000	400	140
8	800	160	56

2,000 + 800 + 400 + 160 + 140 + 56 = 3,556
35. 30 caramel apples

WEEK 8 (pages 19–20)
36. The 6 goes in the middle, as it is an even number that is divisible by 3.
37. 70
38. 3/12 or 1/4
39. 1,605 miles
40. $91.25

WEEK 9 (pages 21–22)
41. 6
42. Division; the array shows a remainder.
43. 63.63
44. 8 and 12
45. 288,000 cookies

WEEK 10 (pages 23–24)
46. $p = 6$
47. $1,000,000 = 10^6$
48. 8 cubes
49. 1/3 + 1/6 = 3/6 or 1/2
50. 4 weeks

WEEK 11 (pages 25–26)

51. 2 (If you multiply the first two digits of the top row, the product is the first two digits of the bottom row, and so on.)
52. 1/4 + 1/3
53. 2-1/2 gallons
54. $4 \times 12 + 36 \div (9 - 8) = 84$
55. 1,320 inches

WEEK 12 (pages 27–28)

56. 1.8 + 2.5 or 1.5 + 2.8
57. 4-1/4 – 1/2 (or 2/4) = 3-3/4
58. 2/8 and 4/16
59. 7 boxes
60. 150 inches; 27 inches longer

WEEK 13 (pages 29–30)

61. 24 more hours
62. $x = 3$
63. 15
64. 3.8 inches
65. 10 aliens with 2 arms and 6 aliens with 4 arms

WEEK 14 (pages 31–32)

66. 2/3
67. $y = 51$
68. 28 tomatoes
69. 132 grapes
70. $568

WEEK 15 (pages 33–34)

71. 6/8 – 1/2
72. 2
73. 12, 16, 8
74. 1,000 pounds
75. 36 minutes

WEEK 16 (pages 35–36)

76. tens place
77. 9
78. Car 2; by 5 miles
79. 32 monkeys
80. AJ is 37 years old, JJ is 31, and TJ is 11.

WEEK 17 (pages 37–38)

81. 0
82. No; the correct equation is $(20 \times 6) + (7 \times 6)$.
83. 250 days
84. 12 6-inch shoes
85. 0.48 ounces

WEEK 18 (pages 39–40)

86. 9
87. False; 6/8 = 3/4 (or 2/3 = 6/9)
88. $x = 4$
89. 6 batches
90. $149.40

WEEK 19 (pages 41–42)

91. 127.2
92. 40; equations may vary.
93. 6
94. 68 frogs
95. 5 times; 1 hour, 10 minutes leftover

WEEK 20 (pages 43–44)

96. 14/5 = 2-4/5
97. Answers will vary. Possible answers: $123.6 - z = 76.9$ or $76.9 + z = 123.6$
98. 49
99. 68; problem-solving methods will vary.
100. 16 pieces

WEEK 21 (pages 45–46)

101. 3.999, 3.997, 3.99, 0.9
102. Possible answer:

103. True
104. 0.14 mph
105. 18 times

WEEK 22 (pages 47–48)
106. 10 snickels
107. 9
108. (6 × 3) + 7 = 5 × 5
109. 17/16 or 1-1/16 feet
110. Roy is 12 years old, and his sister is 6.

WEEK 23 (pages 49–50)
111. 16/32, 32/64, 64/128
112. 14 and 21
113. $q = 18$
114. 30 seconds or 1/2 minute
115. 64 days

WEEK 24 (pages 51–52)
116. 8/3 = 2-2/3
117. 4.03 + 4.4 + .35 = 8.78
118. P = l + l + w + w; A = l × w
119. Answers will vary. Possible answers:
     8/120, 1/15, 4/60
120. $2.85 more

WEEK 25 (pages 53–54)
121. 1/4
122. 0.6

123. 30.1 tiddlywinks
124. 3 cups
125. 6 inches by 6 inches (36 square inches)

WEEK 26 (pages 55–56)
126. 875/1,000 = 7/8
127. 1/2
128. 2,000 zeroes
129. 3.13 inches longer
130. 146 square feet

WEEK 27 (pages 57–58)
131. 10
132. **4**/50 = 16/**200**
133. 4 units by 4 units
134. 28 days
135. 50,000 more legs

WEEK 28 (pages 59–60)
136. 3,195
137. 10,000; greater
138. 5/12 and 1/2
139. 12.4 cm
140. $6.40 an hour

WEEK 29 (pages 61–62)
141.

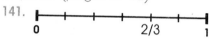

142. 1/9
143. 14,397.3
144.

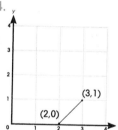

145. l = 12 cm and w = 3 cm

WEEK 30 (pages 63–64)
146. 138, 141; + 3, × 2
147. 8/16
148. 26; it's not a perfect square because
     it is not the product of one number
     multiplied by itself.
149. 112th day
150. 19/24 of the cookie

WEEK 31 (pages 65–66)

151. 0.95
152. 17
153. $V = l \times w \times h$
154. rectangle

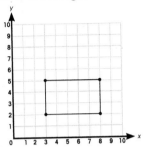

155. 2 squares and 7 triangles

WEEK 32 (pages 67–68)

156. May 6
157. 9
158. 5 units
159. 64

```
            4
           10
           50
   117)7,488
      5,850  ×50
      1,638
      1,170  ×10
        468
        468  ×4
          0
```

160. 2 cups of sugar

WEEK 33 (pages 69–70)

161. 7/8
162. No; the student didn't line up the decimal points properly. The correct answer is 1.83.
163. 1,200 + 280 + 150 + 35 = 1,665

	40	5
30	1,200	150
7	280	35

164. parallelogram

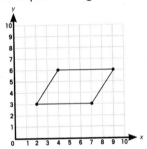

165. 352 cubic feet

WEEK 34 (pages 71–72)

166. 35
167. .00279 (also consider accepting 0.0279)
168. 121, 169
169. 20 minutes
170. 2 square cm

WEEK 35 (pages 73–74)

171. Remove 17 and 42
172. 7
173. Volume should be given in cubic cm.
174. 4 times as great
175. 1 hour, 3 minutes

WEEK 36 (pages 75–76)

176. 1/4 and 3/24
177. half
178. 27 smaller cubes
179. 5th grade
180. You are!